Graphs and Charts

by

Karen Bryant-Mole

Illustrated by

John Yates

Wayland

Notes for teachers and parents

This book investigates the way in which information can be collected and presented. It helps children to choose the most appropriate way of displaying data in chart form and encourages them to look at graphs and charts in the media, paying particular attention to the 'fairness' of such graphs.

Understanding Maths

Adding and Subtracting
Numbers
Measurement
Shape
Graphs and Charts
Multiplying and Dividing

Series editor: Deborah Elliott
Edited by Zoë Hargreaves
Designed by Malcolm Walker
Commissioned photography by Zul Mukhida

First published in 1992 by
Wayland (Publishers) Limited
61 Western Road, Hove,
East Sussex BN3 1JD

British Library Cataloguing in Publication Data
Bryant-Mole, Karen
Graphs and charts. - (Understanding maths)
I. Title II. Series
372.7

ISBN 0 7502 0037 5

Phototypeset by Malcolm Walker
Printed by G. Canale & C.S.p.A., Turin
Bound by Casterman S.A., Belgium

Contents

Pictograms 4
Tally charts 6
Bar charts 8
Bar-line charts 10
Line graphs 12
Grids 14
Pie charts 16
Tree diagrams 18
Decision trees 20
Tables 22
Averages 24
Which type of graph? 26
Misleading graphs 28

Glossary 30
Books to read 31
Index 32

Pictograms

Paul loves vanilla ice-cream. When it was Paul's birthday his mum took Paul and seven of his friends to the park. There was an ice-cream van at the park and his mum bought each of the children an ice-cream. The ice-cream seller had choice of vanilla, chocolate, strawberry and peach flavours.

Each child decided which flavour they would like. Paul's mum ordered 2 vanilla, 3 chocolate, 2 strawberry and 1 peach ice-creams.

Instead of writing this in numbers and words you can show the information in the form of a chart.

This type of chart is called a pictogram or pictograph. The information is shown in pictures.

The chart shows how many children ordered each of the flavours. Each of the ice-creams has its own picture.

Sometimes pictograms are used to represent much larger numbers.

The ice-cream seller sold lots of ice-creams that day.
He sold 73 vanilla, 56 strawberry, 13 peach and 48 chocolate ice-creams.

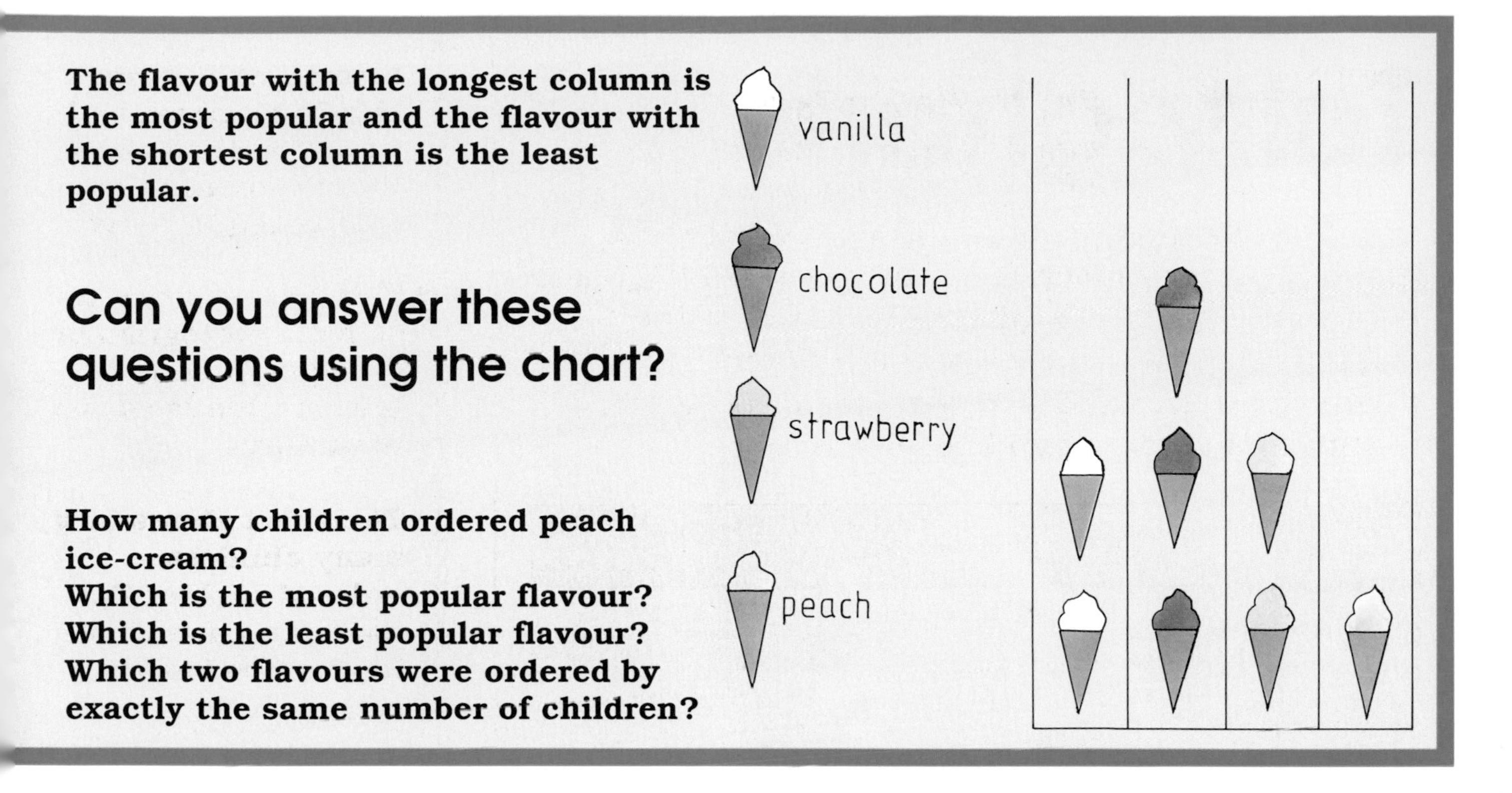

The flavour with the longest column is the most popular and the flavour with the shortest column is the least popular.

Can you answer these questions using the chart?

How many children ordered peach ice-cream?
Which is the most popular flavour?
Which is the least popular flavour?
Which two flavours were ordered by exactly the same number of children?

If he made a pictogram of all those ice-creams, drawing one picture for every ice-cream, he would have to make very tiny drawings or else use a very large piece of paper!
A better way to show this information would be to make every picture of an ice-cream represent 10 ice-creams.

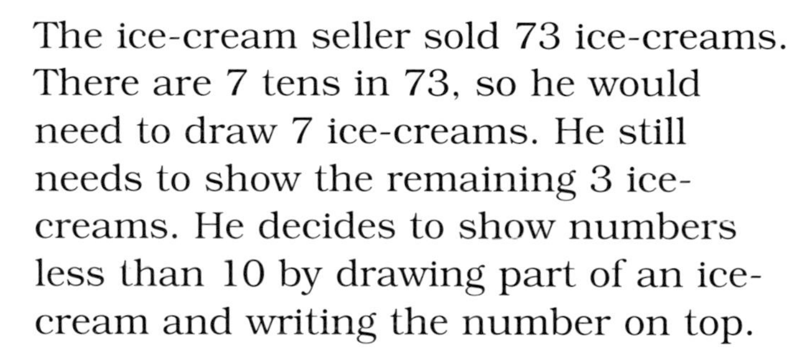

The ice-cream seller sold 73 ice-creams. There are 7 tens in 73, so he would need to draw 7 ice-creams. He still needs to show the remaining 3 ice-creams. He decides to show numbers less than 10 by drawing part of an ice-cream and writing the number on top.

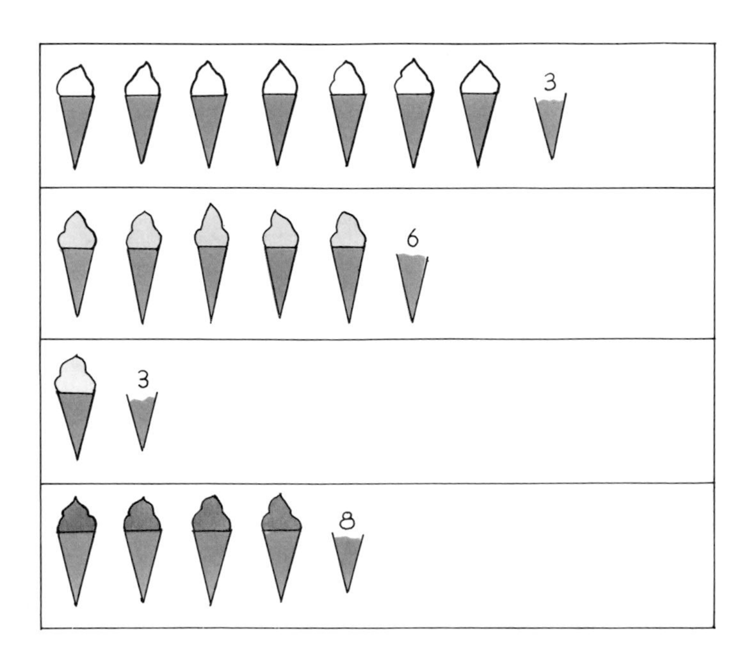

This is what the finished pictogram looks like.

When you make a chart like this always remember to put in a 'key'. The key tells you what the pictures in the chart mean.

The ice-cream seller wants to try out a new toffee flavoured ice-cream, but he still wants to sell just four different flavours of ice-cream. Which flavour do you think he should replace with the toffee?

Tally charts

Simon is conducting a traffic survey. Information collected in a traffic survey can be used in all sorts of ways. It might be used to work out the timing of traffic lights or to estimate how long it will be before the road needs resurfacing.

Simon could make a list of all the vehicles that pass; car, van, car, bus, lorry, car, bicycle, car.

The trouble with this method is that on a busy road he would miss some vehicles in the time it takes to write 'bicycle'. The list is in a jumble. If you wanted to know how many lorries had passed, you would have to go back through the whole list, counting up each lorry.

So, Simon needs to use a method of collecting the information that is quick to write and easy to understand.

This is what he does. It is called a tally chart. Before he begins the survey he writes a list of all the types of vehicles he is likely to see. He adds one more group to the bottom of the list, called 'Other'. This is in case any vehicles pass by that do not fit into the other groups. A tractor, for example, would go in 'Other'.

Traffic passing the station in 5 minutes	
Car	𝍸 𝍸 𝍸 𝍸 IIII
Bus	𝍸
Lorry	𝍸 𝍸 𝍸 II
Bicycle	IIII
Van	𝍸 𝍸
Motor-bike	III
Other	I

As each vehicle passes, Simon draws a small line, or tally mark, in the correct row. The first four lines are drawn next to each other but the fifth tally mark is drawn across the first four.

Look at the row for bicycles. Can you see that four have passed by so far? Now look at the row for buses. Can you see that five buses have passed by?

Grouping the tally marks in fives makes them easier to count. All you have to do is count up in fives until you get to the final group and then count on the last few. You are also less likely to make a mistake. If all the tally marks were written in one long line it would be easy to miss one out or count the same one twice.

Below is the car row written out in groups of five and then written out in one long row.

Try counting them both. The grouped tally marks are much easier to count.

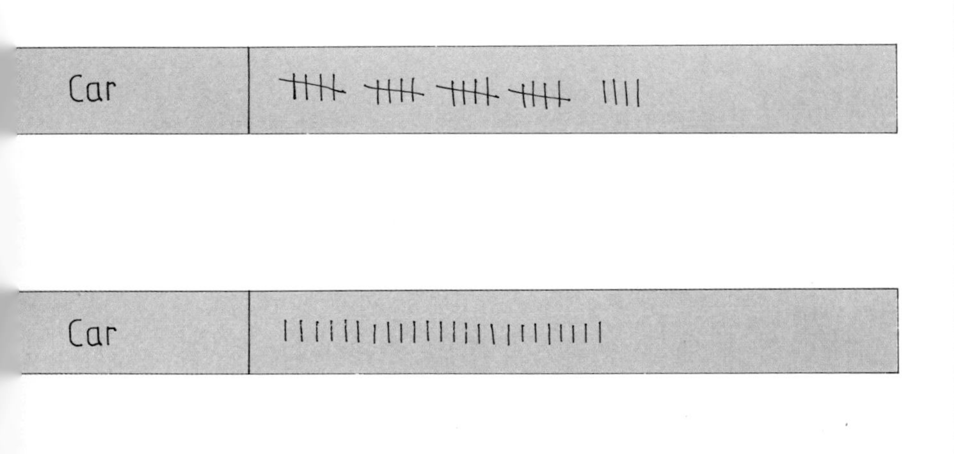

Simon also needs to record the length of time he spent counting the traffic.

is no use recording that 24 cars passed by without lso knowing whether they passed by in 5 minutes or hours.

Try making your own tally chart.

The next time you go on a car journey make a list of some of the things you are likely to see. The things you put on your list will depend on the type of journey. If you are going to be passing through busy towns your list might include supermarkets, traffic lights and taxis. If your journey is going to be in the countryside your list might include horses, cows and tractors.

Decide how long your survey is going to last and fill in your tally chart for that length of time.

Bar charts

When Simon collected his information he used a tally chart. He could just leave it as a tally chart but he wants to show the information in a way that makes it even easier to see what type of vehicle passed by most often.

Simon chooses to make a bar chart of the results.

This is what it looks like.

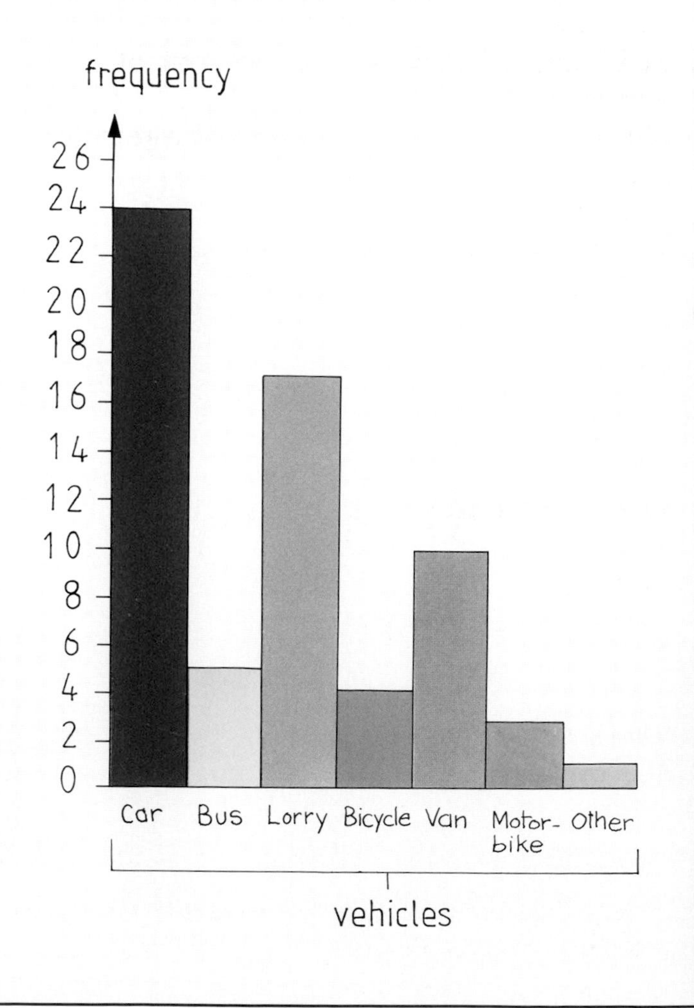

The line along the bottom (or the horizontal axis) lists the names of each group of vehicle in the survey.

The line up the side (or the vertical axis) is used to show how many there were (or the frequency) of each vehicle.

Notice that the vertical axis does not show every single number between 0 and 30. This is because the numbers would be too squashed up.

Look at the bus bar. You will see that the bar stops between the 4 and the 6. The number 5 comes between 4 and 6, so this bar shows that 5 buses passed by.

Compare this chart with the tally chart on the previous page. Can you see that they both show the same information?

Graphs and charts are used to show information in a way that is easy to understand. Other words that you might come across when you are looking at graphs and charts are 'data' and 'statistics'. These are just different words for information that has been collected.

When data is recorded and used, especially when it is used by a computer, it is called 'data processing'.

Bar charts like the one on the opposite page can be used to show many different kinds of information. Sometimes there is just too much information to give each piece of data a separate bar. You have to make each bar represent a group (or range) of data.

These children are playing tiddly-winks.

There are 40 tiddly-winks and they are each trying to get as many as possible into the pot. The highest score they can get is 40 and the lowest is 0. If each score between 0 and 40 had its own bar the chart would have to be very wide indeed.

A clearer way to show the results would be to group the scores so that, for instance, all scores between 0 and 10 form the first bar, all scores between 11 and 20 form the second bar, and so on.

Here is the bar chart showing the results.

What do you notice about the bars?

What does this tell you about the game?

frequency

8
7
6
5
4
3
2
1
0

0-10 11-20 21-30 31-40 scores

Bar-line charts

Laura goes horse-riding every weekend. She also likes to play tennis. Exercise is important for our health. In a survey, 100 people were asked what sort of exercise they had done in the past four weeks. The results of the survey are shown in the graph below. This type of graph is called a bar-line chart. It is like a bar chart but the numbers are represented by a line rather than a wide column.

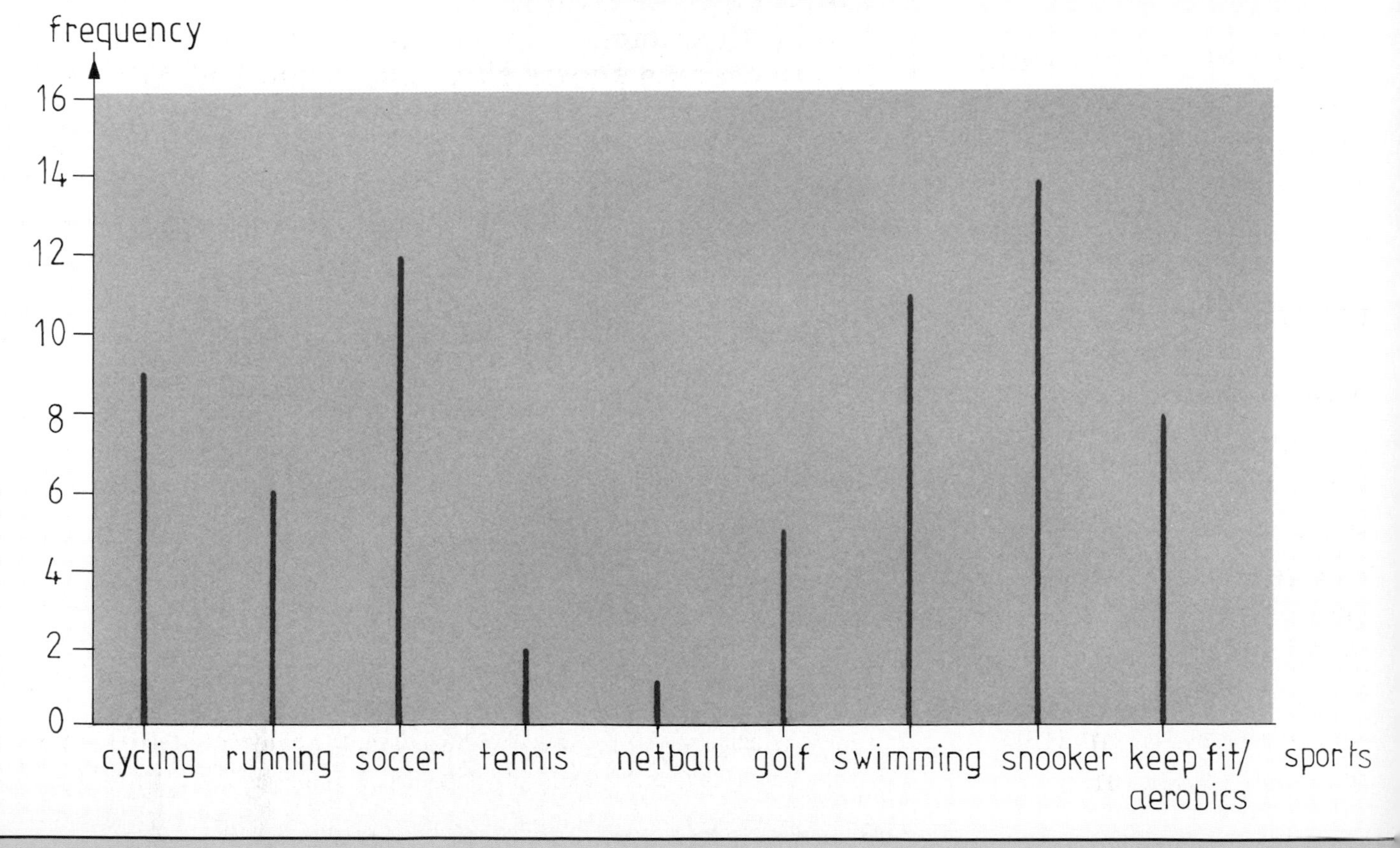

A bar-line graph allows you to show both sets of information on the same graph.

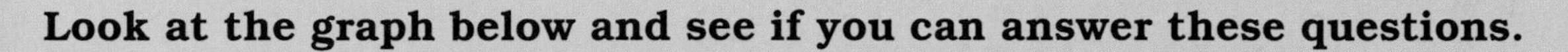

Look at the graph below and see if you can answer these questions.

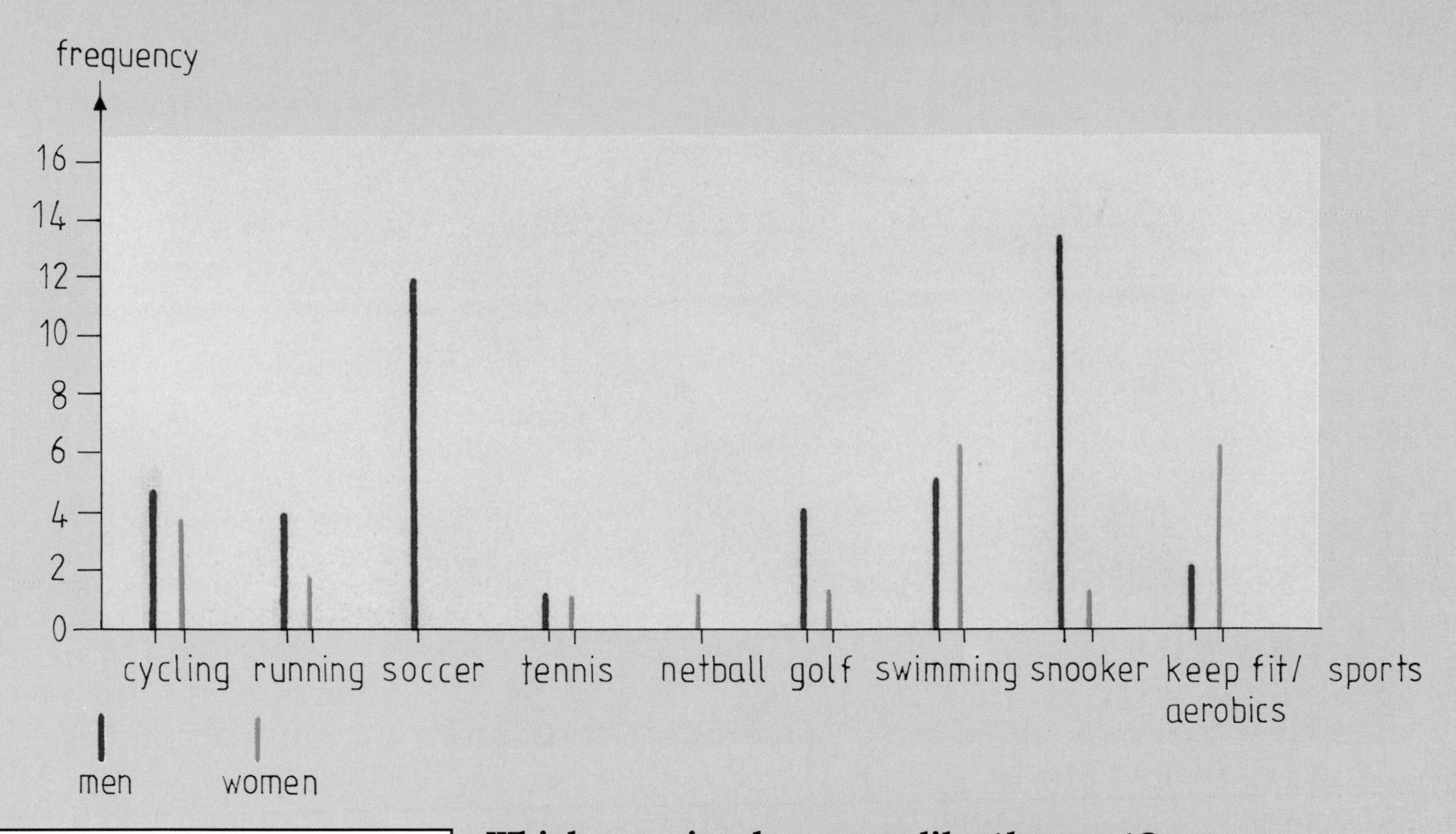

If you look at the graph you can see that snooker is the most popular sport and soccer is the second most popular.
But this does not tell you the whole story.

Although soccer and snooker are increasingly being enjoyed by women, men and women do not always enjoy the same types of exercise. The graph would be more accurate if we split the survey results up into men and women.

Which exercise do women like the most?
Which is the most popular form of exercise for men?
If you owned a sports shop would you buy more pairs of ladies' running shoes than men's?
If you taught aerobics would you run more women's or men's sessions?
How many men in the survey played netball in the last four weeks?

Find out which are the most popular forms of exercise with your friends.
Make a list of activities they have done in the past two weeks and make a tally chart of their answers.
When you have finished your survey make a graph like the one above.

Graphs like these are very useful when you want to compare two sets of information. It is not only boys and girls or men and women that can be compared. This sort of graph is often used to compare survey results from one year to the next, one country with another country or perhaps one make of car with another.

Line graphs

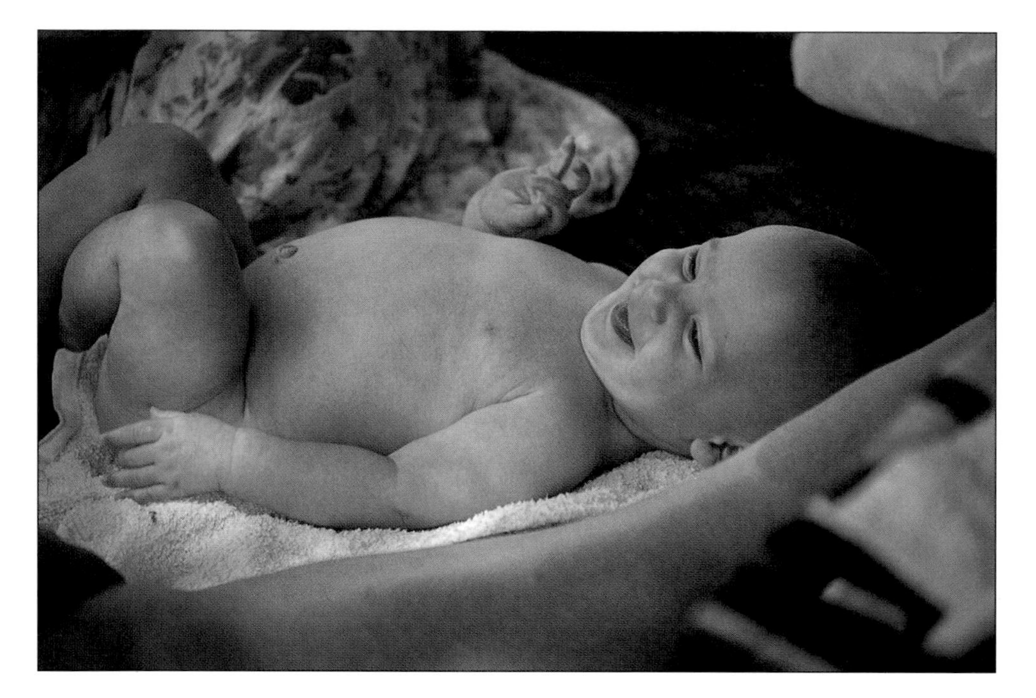

Lucy weighed 3.4 kilograms (kg) when she was born. A month later she weighed 4 kg. Her mother takes her to the clinic every month to be weighed. The doctor makes a note of her weight and makes sure that she is putting on the right amount of weight every month.

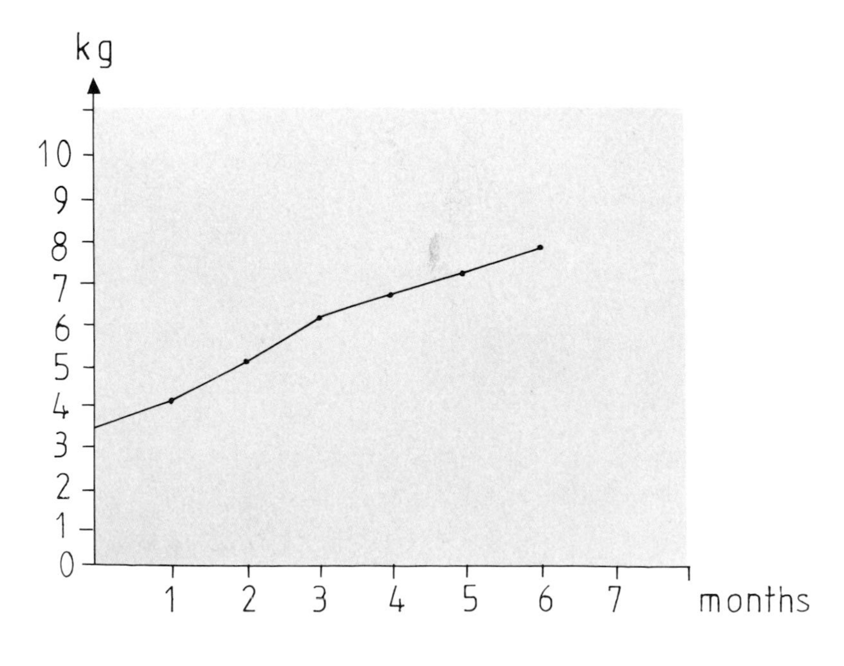

The doctor records the weight in the form of a graph. The graph below shows Lucy's weight gain in the first six months of her life.

This type of graph is called a line graph. Line graphs are often used to show change, such as growth, over a period of time.
Bar charts show the differences between things.
Line graphs usually look at changes in the same thing (in this case Lucy's weight).

The dots on the line graph show the actual measurement taken.

Put your finger on the third dot. Move it downwards until you reach the horizontal axis. This tells you Lucy's age when the measurement was taken.

Put your finger back on the third dot. Move it sideways until you reach the vertical axis. This tells you her weight in kilograms.

When Lucy was two months old she weighed 5 kg.

How much did she weigh when she was four months old?
At what age did she weigh 6 kg?

From looking at the graph it is possible to guess Lucy's weight next month. Do you think it will be 6 kg, 8 kg or 12 kg?

The graph also enables you to estimate Lucy's weight at other times.

If you wanted to know roughly how much Lucy weighed at 3 months you would put your finger on the horizontal axis between the 3 and the 4. Move it up until you get to the line and then move it sideways until you get to the vertical axis.

This will not tell you her exact weight at that time because it was not measured, but it will give you a good idea of what it was likely to be.

Sometimes there are two or more lines on a graph. This shows changes in two or more different things over the same period of time.

The graph below compares the number of t-shirts with the number of jumpers a shop sells over a period of one year.

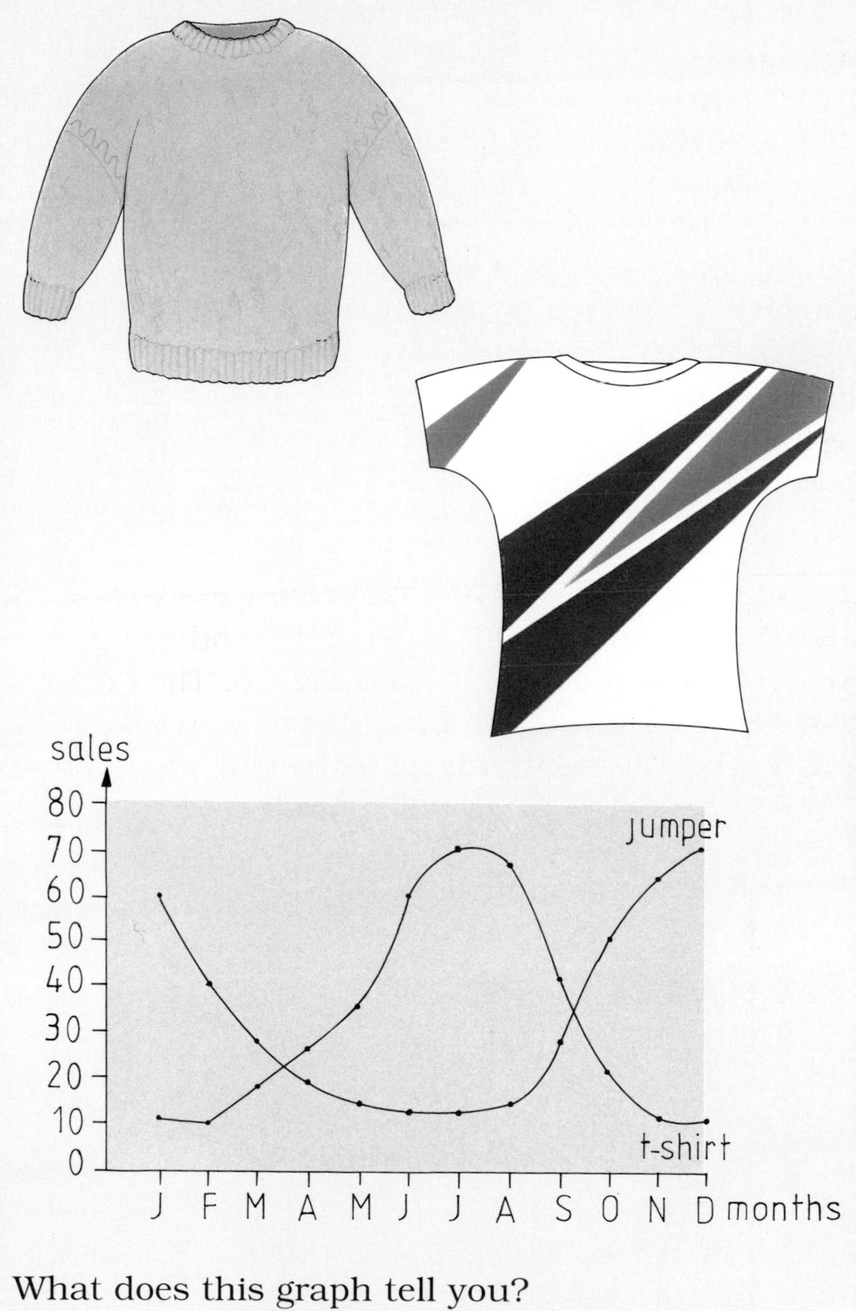

What does this graph tell you?
Think of four different things.
Keep your eyes open for this sort of graph in newspapers and on television.

Grids

James is planning his next trip on his boat. He is using a map to decide which is the best route to take. Maps are drawn on paper that has been marked out in squares. This squared paper is called a grid. Grids help you to work out a position very accurately.

When James is at sea he has to give the coastguard his position every day. Instead of telling the coastguard that he is 'just past the lighthouse' or 'near the harbour', James gives her a grid reference that tells her exactly where he is.

This is how grids work.

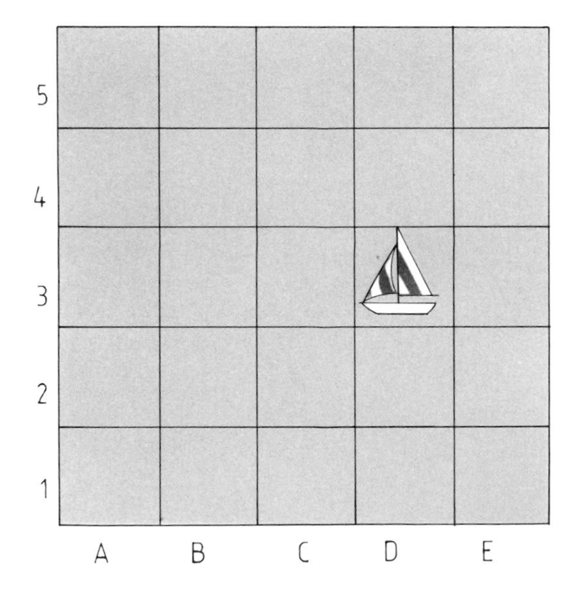

James' boat has been drawn on the map. To work out the position of the boat put your finger on the boat and move it down the column to see which of the letters it is above.

Write this letter down. Now put your finger back on the boat and move it sideways to see which number it is next to. Write this number next to the letter you have already written.

You should have written D3.

It is important to remember to write the letter from the bottom row first and the number from the column at the side second. The first letter (or number) in a grid reference tells you how far along to look and the second number tells you how far up.

Treasure hunt - a game for two players

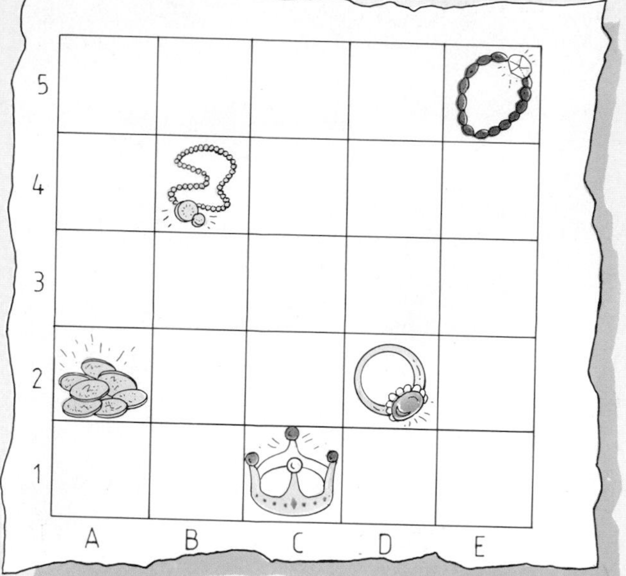

To play this game you will need four sheets of paper.
Each sheet should be marked out in a grid that is five squares long and five squares wide.
The grid should have the letters A to E written along the bottom and numbers 1 to 5 written up the side.
Each player needs two sheets.
On one of these sheets each player should draw in five pieces of treasure anywhere on the grid.

The idea is to see who can find all of the other player's treasure first.
The first grid shows you where your treasure is hidden. Do not show this to your friend.
You will use the second grid to try and work out the position of your friend's treasure.

Player one gives player two a grid reference, e.g A3, and asks whether he has any treasure there.
Player two looks at his or her treasure map. If the answer is 'yes', player one should colour in that square on his or her second sheet of paper. If the answer is 'no', player one should put a cross on that square to remind him or her not to give that grid reference again.
Take it in turns to give each other grid references.

The first person to find all of the other person's treasure is the winner.

Grid pictures

You can use grid references to draw pictures too. Draw another grid four squares high and five squares wide. Colour in the squares shown in these grid references.

A1 A2 A3 A4 B1 B2 B3 C3 C4 D1 D2 D3 E1 E2 E3 E4

What have you drawn?

Pie charts

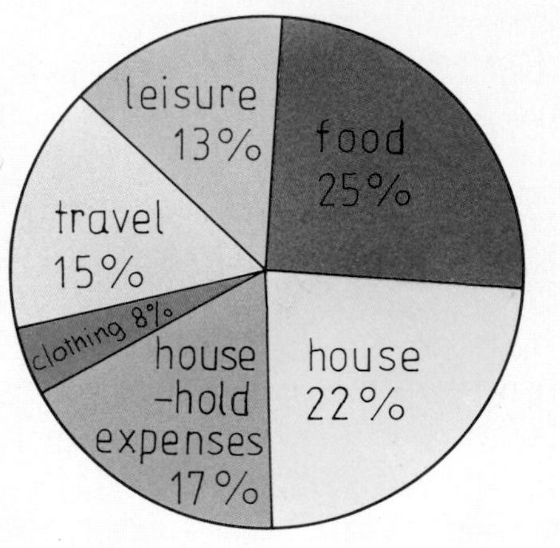

Lisa is doing her weekly shop in the supermarket. She spends a lot of the money she earns on food. The money that you earn is called your income. Lisa spends 25% (per cent) of her income on food. 25% of her income means that for every £100 she earns she spends £25 on food.
She spends 22% on her rent, 17% on household expenses such as electricity and phone bills, 8% on clothing, 15% on travel and 13% on leisure activities such as going to the cinema.

This information can be shown on a chart called a pie chart.
This is the pie chart which shows how Lisa spends her income. It shows you at a glance what she spends most and least on.
It works like this.
Imagine that the circle has been divided up into 100 segments.
Lisa spends 25% of her income on food, so 25 segments are coloured in.

A pie chart like this does not show you how much Lisa earns or how much she spends on each thing.
The pie chart shows you how big or small a share of the whole amount she spends in each area.

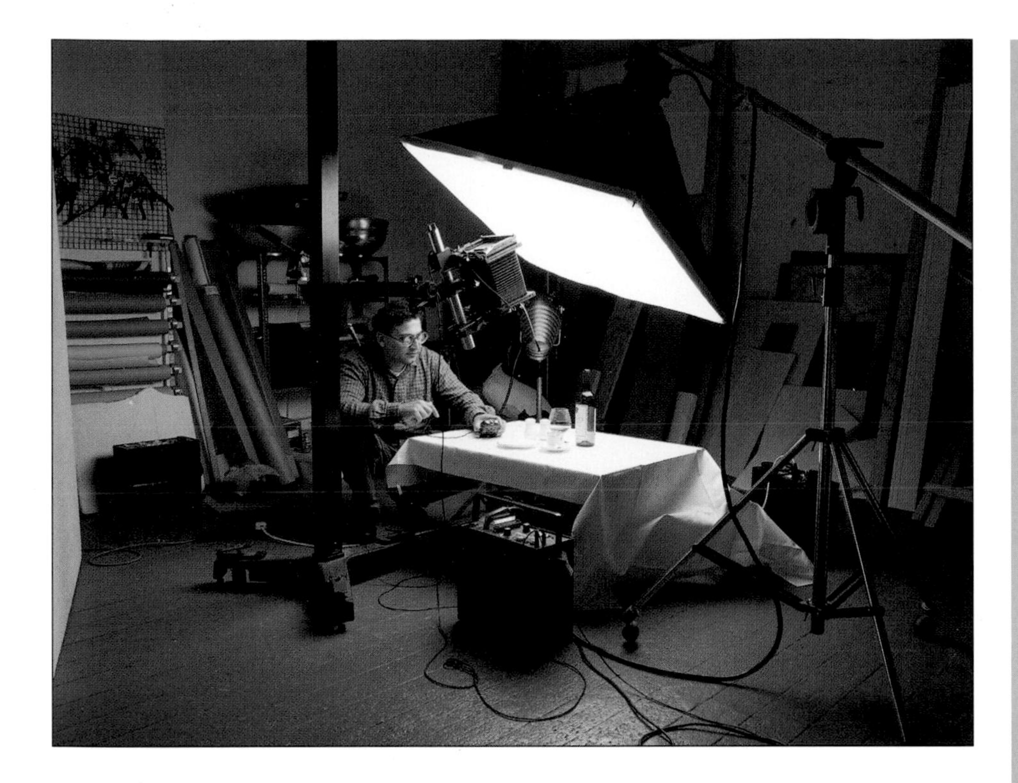

This is Zul, a photographer. On a typical day he spends 8 hours working in his studio, 1 hour travelling to and from work, 2 hours eating his meals, 2 hours watching television, 1 hour playing with his children, 1 hour doing housework, 1 hour reading and 8 hours sleeping.

Working out percentages and dividing circles into 100 segments involves difficult sums but you could make a pie chart of Zul's day without having to do this.

A day is made up of 24 hours. If you divide your circle into 24 segments each segment will represent one hour. This is what Zul's day looks like.

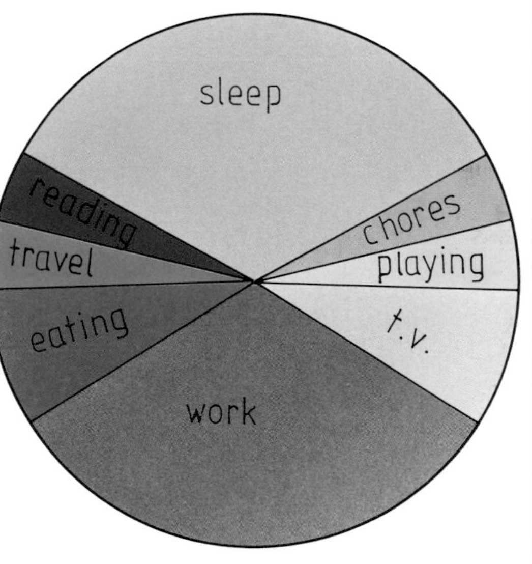

Which two activities take up most of the day?

Make a pie chart of your day

Trace this circle, with its segments, on to a piece of paper.

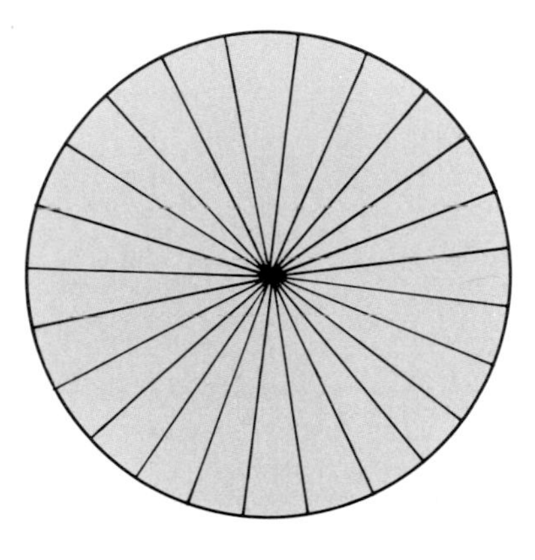

Keep a note of how much time a day you spend doing things like watching television, working at school, eating and playing and make a pie chart like Zul's.

Compare it with Zul's.

Are there things that appear on your chart but don't appear on Zul's?

Do you spend the same amount of time sleeping as Zul?

Tree diagrams

This is a photograph of the Jani family.
There are three generations of the family in the photograph.

At the back, on the left, is Aril. His wife Madhur is standing in front of him. The man holding the baby is called Amita. Amita is the son of Aril and Madhur. The baby is Amita's son. The baby's name is Ashok. The lady standing next to Madhur is Indira. She is the daughter of Aril and Madhur and the sister of Amita. Next to Indira stands Sunita. She is the wife of Amita and the mother of Ashok.

It is much easier to understand this if we draw a diagram.

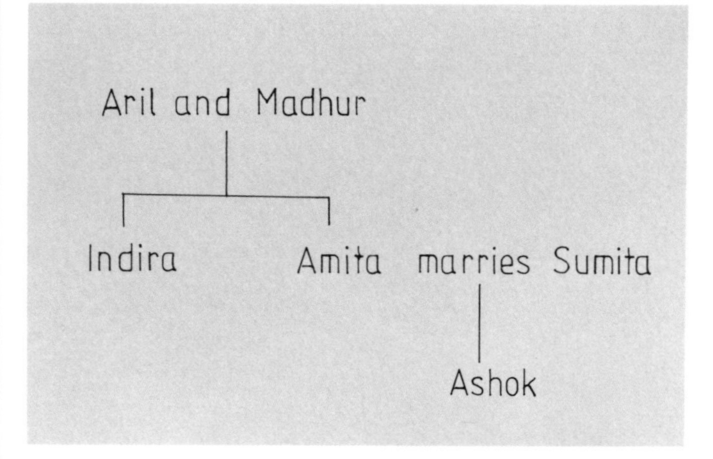

This diagram tells us that Aril and Madhur have two children Indira and Amita. Amita is married to Sunita and they have one child, Ashok.

Who is Ashok's grandmother?

Who is Madhur's daughter?

Who is Indira's sister-in-law?

This type of chart or diagram is called a tree diagram.

They are useful when looking at the relationship between things, animals or people.

Many people are interested in tracing their family history and finding out who married whom and how many children they had. There is a special name for the study of families. It is called genealogy.

Stephen made a tree diagram of his family.

He started with his grandparents, James and Florence Carr and Bert and Margaret Ford. James and Florence had three children, John, Hester and George Carr.

Bert and Margaret had two children, Susan and Catherine Ford. Susan Ford and George Carr got married and they have two children, Stephen and Amy Carr.

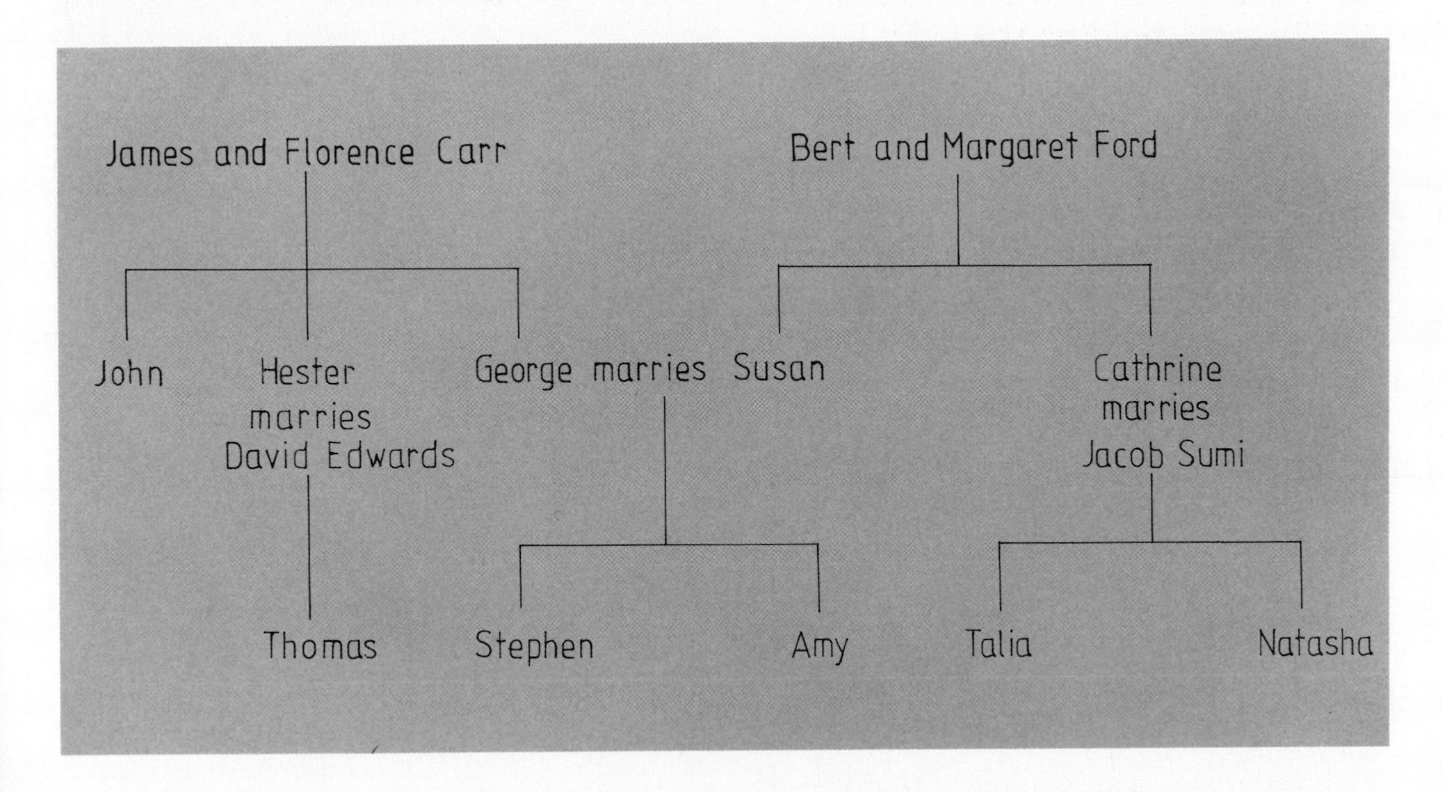

Hester Carr married David Edwards and they have one child, Thomas. John Carr is not married.

Catherine Ford married Jacob Sumi and they have two children, Talia and Natasha.

How many aunts and how many uncles has Stephen?
How many cousins has Stephen?

Make a tree diagram of your family. See if you can find the date of birth of each of the people in your tree and write that in too.

Decision trees

Every week we throw away tonnes of rubbish. Unfortunately, we are using up the earth's valuable resources by using things once and then throwing them away.
One way in which we can help to preserve the earth's resources is to recycle as much material as possible. But what can be recycled?

Here is a chart to help you work out whether or not something can be recycled.
This type of chart is called a decision tree.
It guides you through a series of questions to which you answer 'yes' or 'no', and eventually leads you to a decision. In this case the decision tree tells you what action to take.

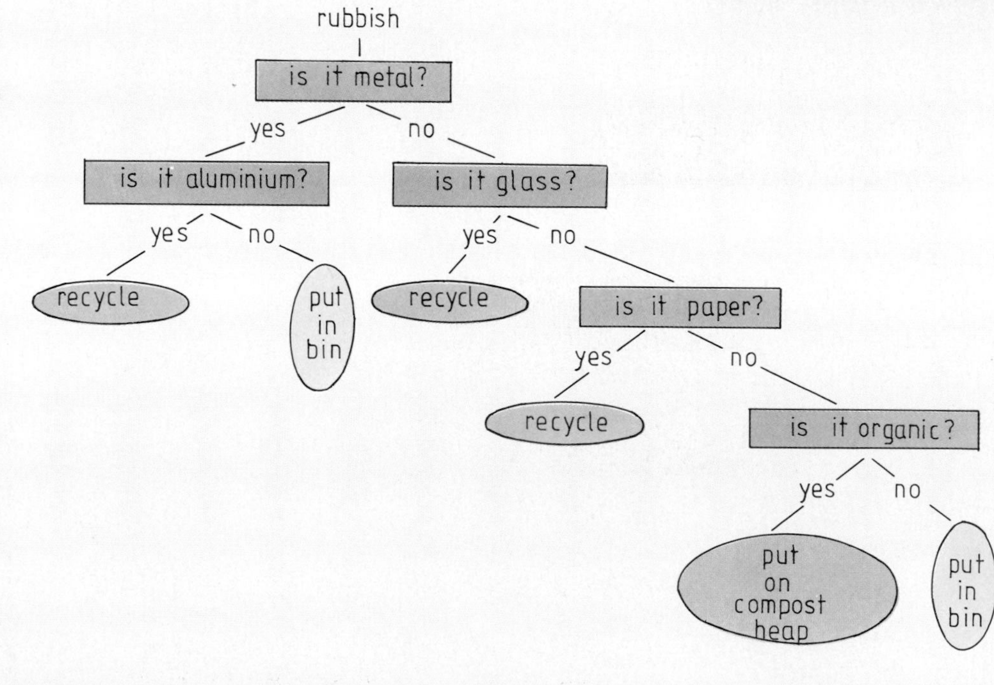

As you can see, not everything can be recycled but you can certainly reduce the amount of rubbish you put out for the refuse collectors every week.

Decision trees are used to sort and classify things.

Try this experiment

Put three teaspoonfuls or 3 x 5 ml of each of these substances into four identical containers; flour, salt, icing sugar, bicarbonate of soda.

Ask a friend to muddle up the containers so that you cannot possibly know which one is which. Tell your friend that you will soon be able to tell him or her what is in each container.

Follow the decision tree to work out what each substance is.

There are three parts to the experiment.
First of all you need to mix one teaspoonful of each substance with some vinegar.
Secondly, you need to mix one teaspoonful of the remaining three substances in some water.
Thirdly, you will need to dissolve one teaspoonful of the last two substances in five teaspoonfuls of water, pour the two mixtures into separate ice cube trays and put them in the freezer for one hour.

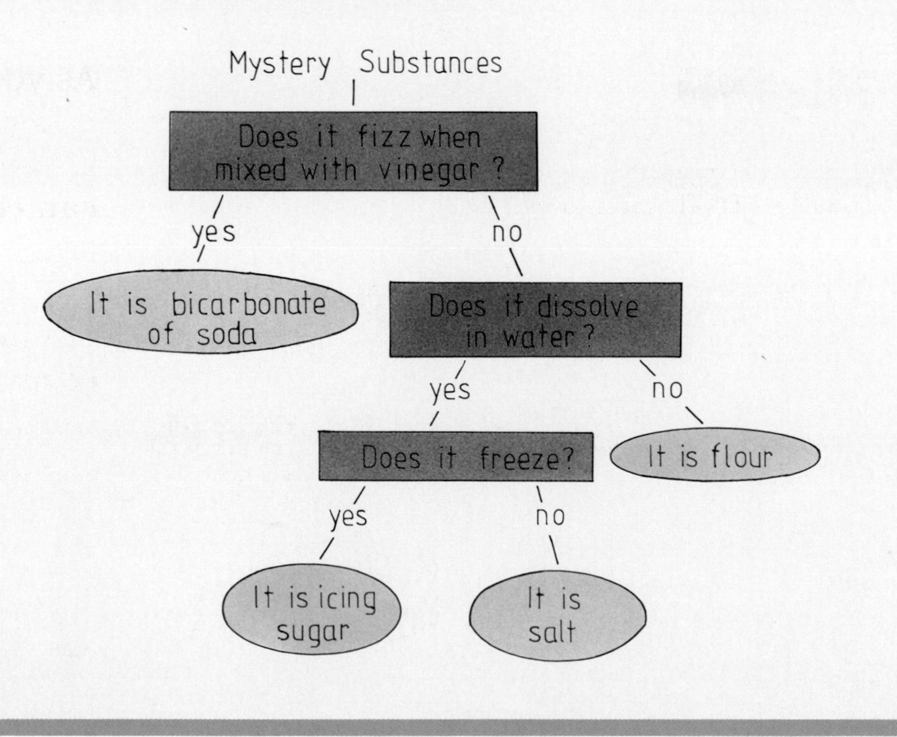

Tables

Information is often printed in the form of a table, for example bus and train timetables.

Northtown	8:10	8:40	9:10	9:40	10:10
Southbridge	8:15	8:45	9:15	9:45	10:15
Oldfield	8:30	9:00	9:30	10:00	10:30
Gatleigh	8:50	9:20	9:50	10:20	10:50
Homestead	9:00	9:30	10:00	10:30	11:00

Look at this bus timetable.
Which bus would you have to catch from Northtown to reach Oldfield by 10:15?

What is the earliest time you could get to Homestead if you could not leave Southbridge before 8:30?

Tables like these show vast amounts of information in a small space.
Tables are not only used to show times. They can be used in many other ways.

Tables are used in sports to compare the results of teams or individuals.

Here is a table showing a season's results for the American football teams in the National Conference Central Division.

How many games did Minnesota win?

Have a games competition with some friends and make a table of the results.

You can play any game in which you can win, lose or tie.

Joel and three of his friends decided to have a hoop-la tournament.

National Conference

Central Division

	W.	L.	T.	Pct.
Minnesota	8	7	1	·531
Green Bay	8	7	1	·531
Detroit	7	9	0	·438
Chicago	7	9	0	·438
Tampa Bay	5	11	0	·313

Every player had to have one game with each of the other players so they all played 3 games of hoop-la.

Can you work out how many games were played altogether? Be careful! The answer is not 12.

Here are the results:
Joel won 2 and lost 1.
Richard tied 1 and lost 2.
Neil lost 1 and tied 2.
Ross won 2 and tied 1.

Put these in the form of a table.

Who do you think is the winner of the tournament?
You will have to think up a way of deciding.
Is it going to be the person with the greatest number of wins? If so, which two people will share the title?
Is it going to be the person with least number of lost games?
Which two people would share the title now?

Sometimes a different number of points are awarded for a win, a tie or a loss. Work out who has the greatest number of points if players are awarded 2 points for a win, 1 point for a tie and 0 points for a loss.
Copy out this table and, using the points system, fill it in putting the players in the correct order, with the winner at the top.

	players	won	lost	tied	points
1					
2					
3					
4					

Look through newspapers to see if you can find any other information presented in a table.

Averages

The Portman family are on holiday. Before they booked their holiday they looked through lots of brochures. Holiday brochures are full of tables.

The cost of the holiday, depending on the hotel you stay in and the date you travel, is usually shown in the form of a table.

Travel arrangements are usually shown in a table too, giving you a choice of flight times, days and airports.

Some of the brochures contain tables showing weather conditions at the various holiday resorts.

Here is the table which shows the weather conditions for the area the Portmans decided to visit; the Costa del Sol in southern Spain.

	April	May	June	July	August	September	October
Average Maximum Temperature	21°c	25°c	27°c	30°c	31°c	29°c	25°c
Average Hours of Sunshine	9	10	11	12	11	9	7

Notice that the table gives the **average** maximum temperature and **average** number of hours of sunshine.
Graphs and charts often make use of averages.

This is how the average temperature in, for example, May is worked out. **The maximum temperature each day in May is recorded.**

1st	2nd	3rd	4th	5th	6th	7th	8th	9th	10th	11th	12th	13th	14th	15th	16th
24°C	26°C	24°C	26°C	27°C	27°C	25°C	24°C	23°C	26°C	24°C	21°C	20°C	21°C	23°C	24°C
17th	**18th**	**19th**	**20th**	**21st**	**22nd**	**23rd**	**24th**	**25th**	**26th**	**27th**	**28th**	**29th**	**30th**	**31st**	
25°C	24°C	24°C	26°C	25°C	25°C	26°C	27°C	27°C	27°C	28°C	28°C	27°C	25°C	26°C	

All the temperatures for each day of the month are added together and the total is then divided by the number of days in the month. This is the sum that tells you the average temperature in May.

775 ÷ 31 = 25

the total of all the temperatures ÷ the number of days in May

This type of average is called the 'mean'.
It is the most widely used sort of average.
There are two other types of average. One is called the 'mode'.
The mode is the number that appears most often in a list of numbers. In this case the mode is 24°C.

The other type of average is called the 'median'.

The median is the middle number in the list, once you have put all the temperatures in order.

Here are the temperatures placed in order.

20°C 21°C 21°C 23°C 23°C 24°C 24°C 24°C 24°C 24°C 24°C 24°C 25°C 25°C
25°C 25°C 25°C 27°C 27°C 27°C 27°C 27°C 27°C 28°C 28°C

There are thirty-one days in May, so the middle number is the sixteenth number. In this case the median and the mean are both the same.
They are both 25°C.

If you look in a newspaper you will usually find a chart showing the maximum temperature the previous day.

Keep a record of the maximum temperature for one month and work out the average mean temperature for that month.

Which type of graph?

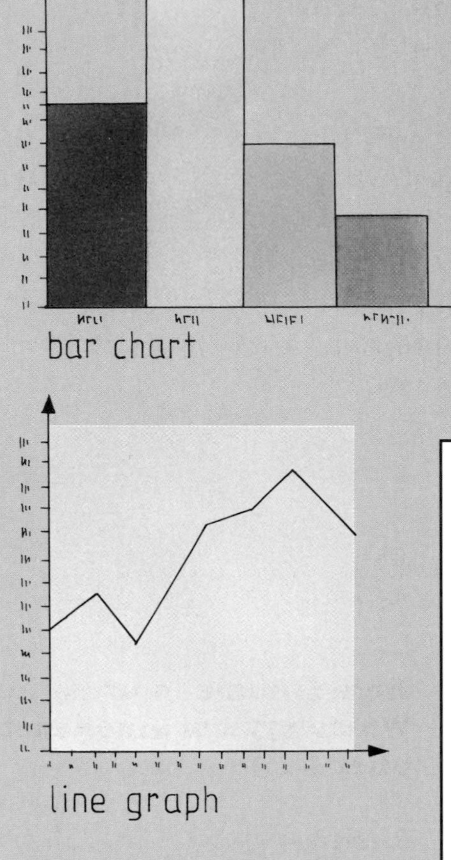

After this chick was born it was weighed every day for three weeks. The farmer kept a chart of the weight increase. Before he started he had to decide which would be the best sort of chart or graph to use. These were the charts he considered:

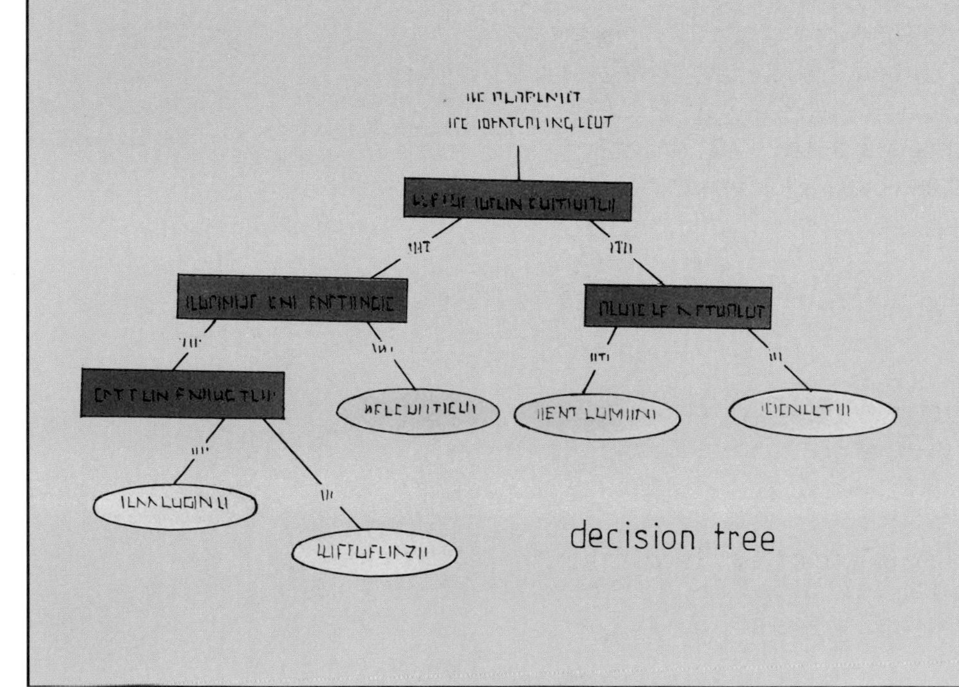

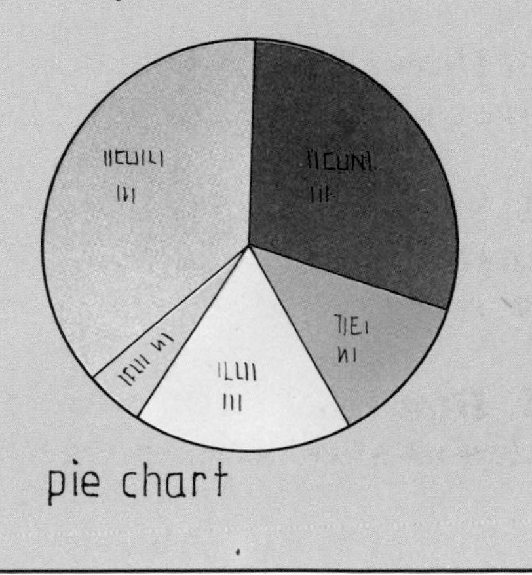

He knew that the tally chart, the pie chart, and the decision tree would be no use at all in showing this sort of information.

Now all he had to do was decide between the bar chart and the line graph. In the end he decided on the line graph because line graphs are very useful for showing continual change, such as growth.

Bar charts are best for separate groups of information, like the number of different types of sandwiches sold in a café in one day.
Tally charts are used to collect and record information quickly.
Pie charts can be used to show 'shares' of a whole amount.
Decision trees guide you through an activity by asking questions to which the answer is 'yes' or 'no'.

Shoe data

Here is some information all about shoes.
Which type of graph or chart would you use for each question?

1. How would you collect information about the shoe colours of people who pass by a busy shop?

2. Natalie had to buy a new pair of shoes. She had to decide whether she wanted indoor or outdoor shoes, what they should be made of, whether they should have buckles, velcro or laces and what colour they should be. Which would be the best graph or chart to use?

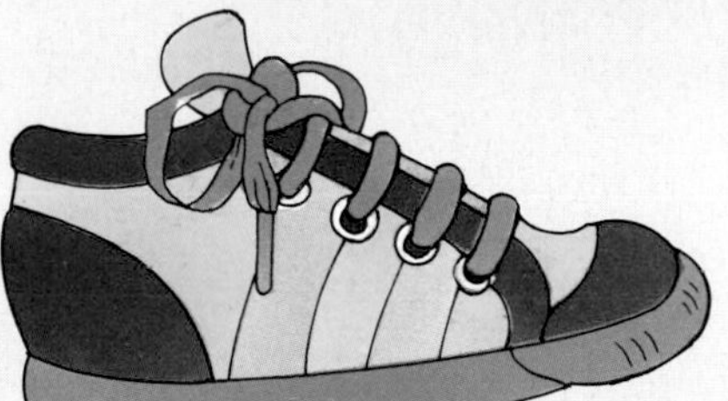

3. How would you show the growth of one child's foot over a period of a year?

4. What sort of graph or chart would you use to show that 70% of a shoe shop's sales were shoes, 10% were slippers and 20% were boots?

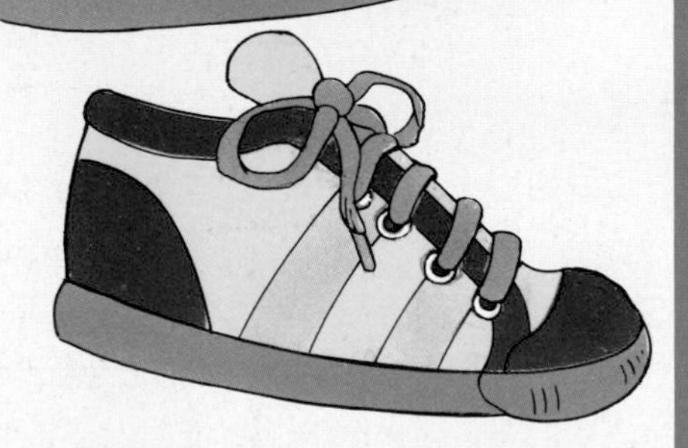

5. How would you show the number of each different size of shoe sold in one day?

Misleading graphs

Vanessa is a market researcher. She collects information about the things people buy.
Manufacturers use this information to find out things, such as why people like their products.
Today is Saturday and Vanessa is asking people whether or not they bought a particular brand of fruit drink that day.
She will ask the same question every Saturday for the next four weeks.

Vanessa gave the information she had collected to the company who make the drink. The drinks company wanted to know whether their new advertisement had increased their drink sales.
This is the graph they drew.

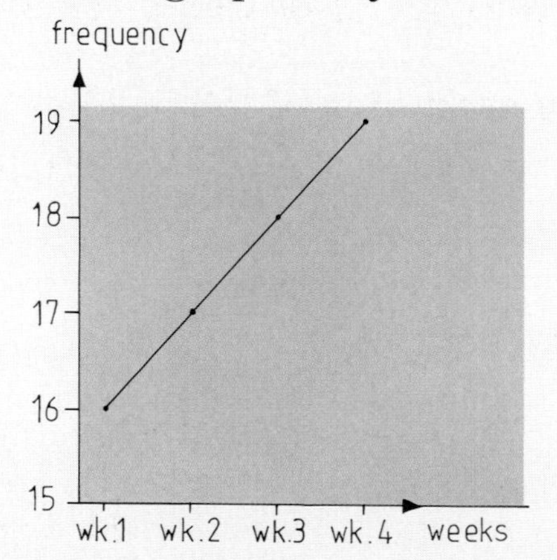

The results seem to show that drink sales have gone from almost nothing to a very large amount. However, look carefully at the graph. Does the axis showing the number of people buying the drink start at 0?

The information shown in the graph is not wrong, but it has been presented in a way that makes the results look better than they really are.
This is how the information should have been shown.

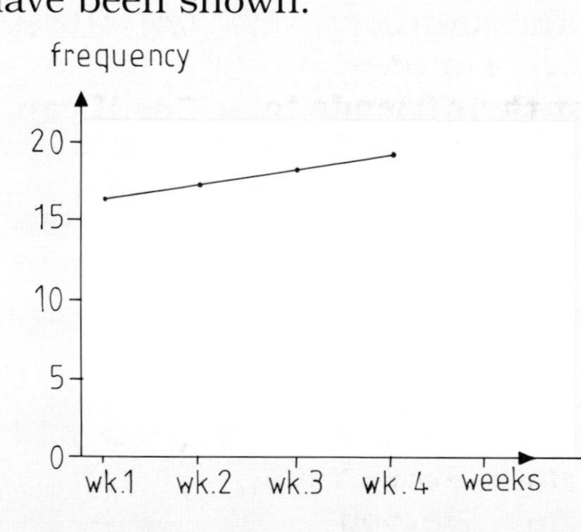

The possible number of people buying the drink starts at 0 and goes beyond the greatest number sold.
Can you see that the results don't look quite so amazing now?

Even that graph does not tell you the complete story. In order to see whether or not the new advertisement is working well, the manufacturers need to compare the sales of their drink with the sales of another similar brand.

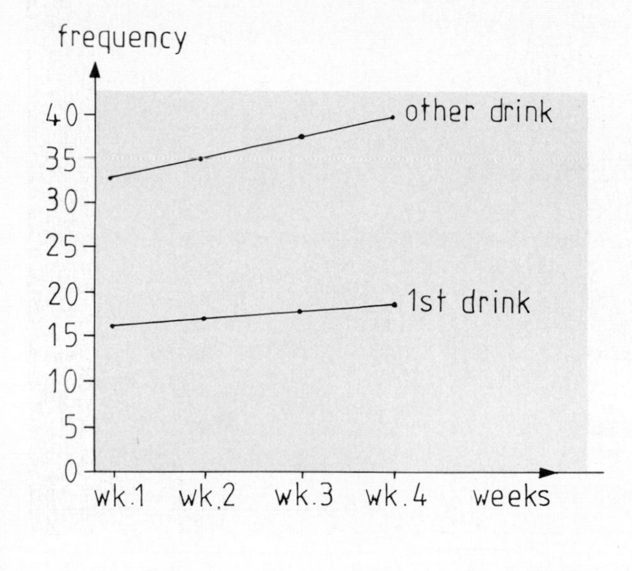

Here is the graph showing that information.

Now the results for the first drink are even less amazing. Although the sales of the drink went up, they are nowhere near the sales of the other drink. What is more, the sales of the other drink went up too, and they did not have an advertisement on television at the time.

Can you think of a reason why the sales of both drinks might have risen?

There are lots of ways in which graphs can be made to look misleading.

Wendy and Andrew conducted a survey amongst their friends to find out which television programmes were the most popular.

They asked them which of the following programmes they had watched on Tuesday; Sports Hour, Story Teller, Action Boy, Cartoon Time, Young News and Art Club.

Here is the chart they made.

It looks as though Action Boy is the most popular programme and no one at all liked Story Teller or Cartoon Time. Now look at the list of programmes on television that day and the times they were shown. Why did no one watch Story Teller or Cartoon Time?

See if you can spot some examples of misleading graphs or charts on television or in newspapers.

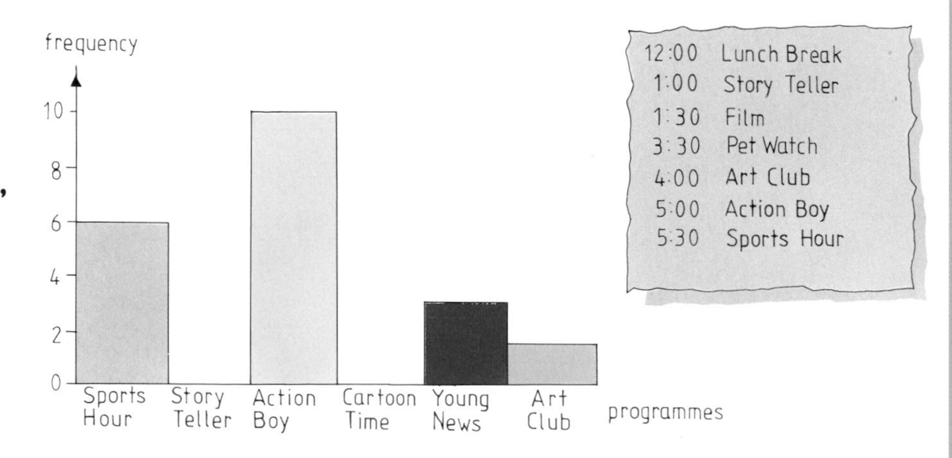

Glossary

aerobics a form of keep-fit

brochure a booklet or leaflet

classify to arrange into groups

estimate to guess the size, quantity or value of something

horizontal from side to side

recycle to use a substance again

survey to collect information

vertical straight up or down

Books to read

Investigating Graphs,
Ed Catherall (Wayland (Publishers) Ltd, 1982)

Usborne Introduction to Maths,
N. Langdon & J. Cook (Usborne)

Photographs in this book were supplied by the following:

Chapel Studios 4, 8, 12, 14, 16, 17, 18, 22, 26, 28;
Wayland Picture Library 6, 7 (Richard Sharpley);
Tony Stone Worldwide 10, 23, 24.

The publisher wishes to thank all those individuals particularly the children and Sally's Model Agency, who participated in and helped with the commissioned photography for this book.

Index

averages 24, 25

bar charts 8, 9, 12, 26, 29
bar-line charts 10, 11

decision trees 20, 21, 26

experiment 21

grids 14, 15
  pictures 15
  references 14, 15

line graphs 12, 13, 26, 28, 29

pictograms 4, 5
pie charts 16,17, 26

tables 22, 23, 24
tally charts 6, 7, 8, 26
tree diagrams 18, 19